TECHNICAL
REPORT

T0162199

Constraints on JP–900 Jet Fuel Production Concepts

James T. Bartis, G. Torrance Flint Jr.

Prepared for the United States Air Force

RAND PROJECT AIR FORCE

The research described in this report was sponsored by the United States Air Force under Contract FA7014-06-C-0001. Further information may be obtained from the Strategic Planning Division, Directorate of Plans, Hq USAF.

Library of Congress Cataloging-in-Publication Data

Bartis, James T., 1945-
  Constraints on JP-900 jet fuel production concepts / James T. Bartis, G. Torrance Flint, Jr.
    p. cm.
  Includes bibliographical references.
  ISBN 978-0-8330-4144-9 (pbk. : alk. paper)
  1. Coal liquefaction. 2. Jet planes—Fuel. 3. Petroleum coke. I. Flint, G. Torrance. II. Title.

TP343.B327 2007
665.5'3825—dc22

                                                                    2007013242

The RAND Corporation is a nonprofit research organization providing objective analysis and effective solutions that address the challenges facing the public and private sectors around the world. RAND's publications do not necessarily reflect the opinions of its research clients and sponsors.

RAND® is a registered trademark.

Published 2007 by the RAND Corporation
1776 Main Street, P.O. Box 2138, Santa Monica, CA 90407-2138
1200 South Hayes Street, Arlington, VA 22202-5050
4570 Fifth Avenue, Suite 600, Pittsburgh, PA 15213-2665
RAND URL: http://www.rand.org/
To order RAND documents or to obtain additional information, contact
Distribution Services: Telephone: (310) 451-7002;
Fax: (310) 451-6915; Email: order@rand.org

Preface

As part of its fuel research portfolio, the Air Force has invested in the investigation and development of processes for producing jet fuel from coal or coal-derived products. Research in this area has progressed far enough that decisions need to be made regarding whether additional investments should be directed toward scaling up the process and conducting large-scale fuel tests.

To better understand the benefits of such additional investments, RAND Project AIR FORCE examined constraints on the commercial viability of two processes under development at Pennsylvania State University.

The research reported here was sponsored by the Deputy Chief of Staff for Logistics, Installations and Mission Support, Headquarters, United States Air Force, in coordination with the Air Force Research Laboratory. This work was part of a larger study, "Unconventional Fuels: Strategic and Program Options," which is being conducted within the Resource Management Program of RAND Project AIR FORCE.

RAND Project AIR FORCE

RAND Project AIR FORCE (PAF), a division of the RAND Corporation, is the U.S. Air Force's federally funded research and development center for studies and analyses. PAF provides the Air Force with independent analyses of policy alternatives affecting the development, employment, combat readiness, and support of current and future aerospace forces. Research is conducted in four programs: Aerospace Force Development; Manpower, Personnel, and Training; Resource Management; and Strategy and Doctrine.

Additional information about PAF is available on our Web site at http://www.rand.org/paf.

Contents

Summary

Researchers at the Energy Institute of Pennsylvania State University (Penn State) are conducting research on producing jet fuel by coprocessing coal or coal-derived products with low-value liquid intermediates produced during petroleum refining. To date, most of this research effort has focused on a coal-tar blending process. Penn State currently plans to build a one-barrel-per-day pilot plant and produce 100 barrels of product to be delivered to and tested by the Air Force Research Laboratory.

Recognizing the limited availability of the coal-tar derived liquids used in the coal-tar blending process, the Penn State research team has recently shifted its attention to a co-coking process, in which a mixture of solid coal and a refinery intermediate, decant oil, is used to produce a combination of liquid fuels and coke.

The findings and recommendations of the RAND Corporation review of these two processes are as follows:

Finding 1

Our review of Penn State's work on JP-900 revealed a research team with considerable expertise in coal pyrolysis. Coal pyrolysis as a means of producing liquid fuels was studied extensively during the first half of the 20th century, but few research teams today have the expertise to address how pyrolysis might be exploited and combined with other liquid-fuel production approaches, including coal and biomass liquefaction via Fischer-Tropsch synthesis, and with other biomass routes to liquid-fuel production. (See pp. 8–9.)

Recommendation
Consider supporting laboratory research and engineering analyses focusing on identifying possible opportunities by which pyrolysis can significantly improve the energy efficiency and costs of producing liquid fuels from coal and/or biomass.

Finding 2

The limited availability of coal tar seriously impedes the ultimate production potential of the Penn State coal-tar blending process. At most, successful development would produce only a few thousand barrels per day of jet fuel. The net displacement of imported oil would be even smaller (pp. 3–4).

Recommendation

Cease all research directed toward developing this process, and cease testing fuels produced from this process. This includes terminating the planned pilot-scale operations at the Harmarville, Pennsylvania research site (p. 8).

Finding 3

For the Penn State co-coking process, the limited availability of decant oil and the limited markets and high quality specifications for premium coke will limit liquid-fuel production to less than 140,000 barrels per day, only a portion of which would likely be suitable for use as a jet fuel. Less than 8,000 barrels per day of this production would be attributed to coal. The net increase in U.S. coal production due to developing the co-coking process would be negligible, about 2 million tons per year, and this estimate assumes that process economics are favorable and that one-half of U.S. decant oil production can be diverted to co-coking (pp. 5–7).

Recommendation

Cease all research directed toward developing (including product testing) any co-coking process concept that depends on large amounts of decant oil for the production of jet fuel. If work is to continue on co-coking, it should be limited to fundamental research investigating the feasibility of co-coking concepts that use feedstocks that are at least an order of magnitude more abundant than decant oil and produce higher liquid and lower coke yields.

CHAPTER ONE

Introduction

Background

Under support from the U.S. Air Force Office of Scientific Research, researchers at the Energy Institute of Pennsylvania State University (Penn State) have conducted research on producing jet fuel by coprocessing coal or a coal-derived product with low-value liquid intermediates produced during petroleum refining. The original objective of this research and development (R&D) effort was to produce a fuel that could serve not only as a source of propulsion energy, but also as a heat-sink, especially for cooling engine components, suitable for use in advanced military aircraft. This fuel was given the designation JP-900 by the Penn State researchers, in recognition of its stability at 900° F.

Under the direction of Professor Harold Schobert, the Penn State research team has developed two process concepts for producing JP-900. To date, most of Penn State's R&D has been directed toward a *coal-tar blending* process, which centers on the hydrogenation of a mixture of a coal tar derivative (specifically, refined chemical oil) and a petroleum refinery intermediate (specifically, light cycle oil). More recently, attention has been directed toward a *co-coking* process, in which a mixture of solid coal and another refinery intermediate, decant oil, is used to produce a combination of liquid fuels and coke.

Both the Air Force and the Penn State research team recognize that fuel production from the coal-tar blending process may be constrained because of the limited availability of coal tar. The Air Force is also concerned that fuel production from the co-coking process may be constrained by the limited availability of decant oil and the limited marketability, as high-value products, of the coke co-products. The Air Force has requested that the RAND Corporation address these issues and make recommendations regarding future directions for this research.

Technical Approach

To understand relevant process concepts, development history, and future research plans, RAND researchers reviewed research reports, briefings, and technical papers provided by the Penn State research team. They also met and held discussions with key members of the Penn State research team and with Air Force Research Laboratory personnel familiar with the project.

The availability of refined chemical oil and decant oil and the potential market for high-value coke products were established based on publicly available information and interviews with petrochemical and coke production consultants and industry members.

CHAPTER TWO

Analysis of Production Constraints

Production Constraints on the Coal-Tar Blending Process

RAND finds that production of jet fuel from the coal-tar blending process will be seriously constrained by the limited availability of coal tar in the United States.

Coal tar is a by-product of the traditional (i.e., chemical recovery) coke ovens used to prepare metallurgical coke. Between 30 and 45 liters (9 to 12 gallons) of coal tar is produced for each ton of coal processed in a chemical-recovery coke oven.[1] With further processing, each gallon of coal tar produces about 0.2 gallons of refined chemical oil.[2] Combining these factors, each ton of coal processed in a chemical recovery coking oven produces approximately 2 gallons of refined chemical oil.

In 2005, coke plants consumed 23 million short tons of coal.[3] Assuming that all this coal went to chemical-recovery coke ovens, we calculate the maximum U.S. production of refined chemical oil to be 46 million gallons per year, or roughly 3,000 barrels per day. This is a maximum, however, because all new coke plants producing metallurgical coke are based on nonrecovery ovens that do not produce chemical by-products. Moreover, not all of this production would be in the middle distillate range and be suitable for use as a jet fuel.

In addition, refined chemical oil produced in the United States is being used for productive purposes, such as manufacturing carbon black, creosote, and certain chemicals. The diversion of refined chemical oil for fuel production would require these current applications use higher-cost alternatives, most likely based on petroleum.

The preceding analysis indicates that the Penn State coal-tar blending process, if fully developed and commercialized, would displace no more than 1 to 2 percent of the 234,000 barrels per day of the petroleum-derived jet fuel the Department of Defense currently con-

---

[1] John T. Baron, Charles E. Kraynik, and Robert H. Wombles, "Strategies for a Declining North American Coal Tar Supply," in M. Sahoo and C. Fradet, eds., *Light Metals 1998 Métaux Légers,* Montréal, Que.: The Metallurgical Society of Canadian Institute of Mining, Metallurgy and Petroleum, August 1998.

[2] Kevin J. Fitzgerald, Vice President and General Manager, Carbon Materials and Chemicals, Koppers Inc., personal communication, November 3, 2006.

[3] Energy Information Administration, *Annual Energy Review 2005,* Washington, D.C.: U.S. Department of Energy, DOE/EIA-0384(2005), 2006a.

sumes.[4] At the national level, crude oil imports would not likely be affected because applications that currently depend on refined chemical oil would have to shift to petroleum-based products.

Fuels Production from the Co-Coking Process

The co-coking process is in a much earlier stage of development than is the coal-tar blending process. In co-coking, solid coal is mixed with decant oil, and the mixture is sent to a delayed coker, which produces significant amounts of coke, as well as fluid products that can be further treated (fractionated and hydrogenated) to produce useful transportation fuels, including jet fuel. The Penn State research team is attempting to produce a premium coke as a coproduct, with the objective of improving the overall economics and commercial viability of the co-coking process.

At its current stage of development, the Penn State co-coking process begins with a 20 percent coal, 80-percent decant oil slurry by weight consisting of cleaned finely ground coal and decant oil. This mixture is fed to the delayed coker.[5] Thirty percent of the coker output consists of coke. The remaining 70 percent consists of fluids (mostly hydrocarbon liquids, about 10 percent gases) that may be further processed to produce liquid fuels, a fraction of which would be potentially useful as a replacement for or blending stock with JP-8 jet fuel. A schematic showing the mass balance around the delayed coker is shown in Figure 1.

Figure 1 shows that the net effect of adding 2 tons of coal to 8 tons of decant oil is to increase the coke yield by 1.8 tons. The fluid product yield is increased by 10 percent of the

Figure 1
Estimated Mass Balance Around the Delayed Coker

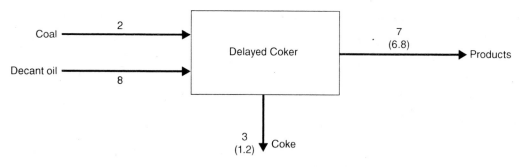

NOTE: Estimated mass balance around the delayed coker. The numbers within parenthesis show the mass balance if only 8 mass units of decant oil and no coal enter the delayed coker.

RAND TR465-1

---

[4]   Energy Information Administration, 2006a, reports annual Department of Defense consumption of 485 trillion British thermal units per year of jet fuel. Using that publication's conversion factor of 5.67 million British thermal units per barrel, this reported quantity is equivalent to approximately 234,000 barrels per day of jet fuel.

[5]   Carolyn Clifford, Energy Institute, Pennsylvania State University, personal communication, November 2006.

input coal mass, (0.2 tons). Moreover, because the energy density of coke is somewhat higher than that of coal, the net energy gain in the product stream may be even less. Given the present state of process development, further research may be able to shift the mass balance toward a greater liquid fraction.[6] Otherwise, it may be more appropriate to consider the co-coking process to be a means of producing a petroleum-like coke from coal, with a small fraction of liquid by-product.

While the addition of coal may not significantly affect the net energy content of the fluid products leaving the delayed coker, the Penn State research team reports that the composition of the products is modified and contains a greater fraction of hydrocarbons that possess the ring structures that are thought to contribute to greater stability and enhanced combustion characteristics at higher temperatures. Thus, the products might, after further processing, yield a jet fuel with improved thermal stability and combustion characteristics.

The challenge for any new process for producing liquid fuels from coal is commercial viability. The commercial viability of the co-coking process depends on whether the change in the value of the product mix leaving a refinery after being modified for co-coking is sufficient to offset the capital investments and operating costs required for such modifications. For example, new equipment (and space at the refinery) will be required for coal delivery, handling, pulverization, and storage. Also, it may be necessary to modify or add to existing refinery equipment, including delayed cokers, fractionation columns, hydrotreaters, hydrogen production units, and product storage. The feasibility of this investment is further clouded by the fairly small amount of decant oil produced at any single refinery, as discussed in the next section.[7] The Penn State research team has not conducted the engineering analyses required to establish that the value of the product mix from a modified refinery will adequately offset the anticipated capital and operating costs for the modification.

Production Constraints on the Co-Coking Process

There are two production constraints on the co-coking process: the availability of decant oil and the market for the additional coke produced.

The Availability of Decant Oil

Decant oil is a by-product of the fluidized catalytic cracking (FCC) units that are major processing operations refiners use to convert low-grade refining intermediates into more-valuable gasoline and middle distillates. In the United States, FCC capacity is about one-third of crude oil capacity. Typically, decant oil represents about 3 to 6 percent of the total liquid product volume leaving an FCC. Recent decant oil production in the United States is reported to be

---

[6] Using a hydrotreated decant oil, Penn State reports that it has obtained liquid fractions as high as 76 percent (Harold H. Schobert, letter, December 12, 2006). However, this approach introduces additional costs for hydrotreating the decant oil. In general, hydrotreated decant is used only for producing needle coke.

[7] For example, a large refinery producing hundreds of thousands of barrels per day of petroleum products produces only a few thousand barrels per day of decant oil. If all of that decant oil were to be dedicated to co-coking, the refinery's net coal use would be only a few hundred tons per day.

15.6 million tons per year,[8] the energy equivalent of about 250,000 barrels per day of crude oil.

The typical refinery in the United States produces less than 5,000 barrels per day of decant oil. Only two refineries, both in the Gulf area, are known to produce between 10,000 and 15,000 barrels per day, and none reportedly produce more.[9] Currently, only a small fraction, about 5 percent, of decant oil is sent to delayed coking units. The largest applications are blending with fuel oil to reduce viscosity and manufacturing carbon black, especially for automotive tire production.

The level of production of decant oil at U.S. refineries and the need to maintain decant oil production for current markets present major constraints on the use of coal by the co-coking process and therefore its potential to produce jet fuel. Assuming that about one-half of annual U.S. decant oil production (8 million tons per year) could be diverted to the co-coking process, the net annual coal consumption would be only 2 million tons per year,[10] and the net fluids yield would be less than 140,000 barrels per day. Assuming that one-half of the total fluids yield would be middle distillates suitable for further upgrading to jet fuel, we calculate an upper bound of 70,000 barrels per day of jet fuel production.

Markets for Petroleum Coke

Since a major by-product of the Penn State co-coking process is coke, the economic viability of this process depends on how much coke is produced and on the market value of and demand for that coke. This analysis assumes that the process will produce a premium quality coke that will sell at prices well above steam coal prices. However, at the current stage of process research, the quality and market suitability of the coproduct coke has not yet been established.

Petroleum coke is produced in refineries as a by-product or, in some cases, coproduct of upgrading the heaviest petroleum fractions to lighter products. During 2005, petroleum refineries in the United States produced 218 million barrels of marketable petroleum coke.[11] According to the Energy Information Administration, about 65 percent of this coke is used as fuel, either domestically or abroad, primarily for electric power generation or in cement kilns. In general, fuel-grade coke is high in sulfur and metals and has a limited domestic market. It is thus a low-value product that is often sold at zero or negative net revenue to the refinery.[12]

---

[8]   Vincent J. Guercio, "Slurry Oils Supply and Demand," paper delivered at the 2004 Fuel Oil/Energy Buyers Conference, Miami Beach, Fla., CTC International, October 2004.

[9]   Vincent J. Guercio, petrochemical consultant, CTC International, personal communication, January 10, 2007.

[10]   This estimate is based on the mass balance shown in Figure 1, which shows 2 mass units of coal use for each 8 mass units of decant oil use.

[11]   Petroleum coke is reported in terms of barrels and tons. The conversion is 5 barrels (42 gallons) per short ton of coke. The Energy Information Administration assumes an average heating value of 6.024 million British thermal units per barrel for coke from petroleum. Energy Information Administration, *Petroleum Supply Annual 2005*, Washington, D.C.: U.S. Department of Energy, DOE/EIA-0340(2005)/1, 2006b.

[12]   D.J. Peterson and Sergej Mahnovski, *New Forces at Work in Refining: Industry Views of Critical Business and Operations Trends*, Santa Monica, Calif.: RAND Corporation, MR-1707-NETL, 2003.

Nearly all the remaining 35 percent, roughly 75 million barrels per year, of petroleum coke is sponge coke that, when calcined, is used in the manufacture of aluminum anodes, furnace electrodes and liners, and shaped graphite products. It is appropriate to assume that calcined sponge coke suitable for these applications will be much more valuable than the coal that would be used in the co-coking process Penn State is developing. A very small amount of petroleum coke production, well under 5 million barrels per year,[13] is directed toward needle coke, which is a premium product used in the manufacture of high-performance graphite electrodes, such as those required in steel arc furnaces.[14]

At the current stage of the development of the co-coking process, the quality of the coke leaving the delayed coker is uncertain. For the purposes of analysis, we assume that this coke is of sufficient quality to displace one-third of current nonfuel demands for petroleum coke in the United States—25 million barrels per year or equivalently, 5 million tons per year.

Given the mass balance shown in Figure 1, in producing 5 million tons of coke, the delayed coking units would also produce nearly 12 million tons of fluids annually. Assuming a yield of 7 barrels per ton of these fluids, the maximum production of liquid fuels from the co-coking process is 84 million barrels annually, or about 230,000 barrels per day. However, it is reasonable to assume that, after refining, roughly half this amount could be middle distillates appropriate for use in jet turbine engines.

Given the mass balance shown in Figure 1 and the energy balance considerations discussed in the previous section, at most only a few percent (i.e., less than 5 percent) of the liquids produced by the co-coking process can be attributed to coal. However, the coal fed into the delayed coker could have a disproportionate influence on the chemical composition of the fluid products leaving the coker.

If the economic viability of the co-coking process requires coproduction of needle coke, the relatively small market for needle coke would severely constrain the amount of jet fuel that could be coproduced. For example, under the assumption that the process could produce 1 million tons per year of needle coke (which is well beyond current U.S. demand), the maximum production of jet fuels would be less than 25,000 barrels per day.

The disposition of the petroleum coke that the Penn State co-coking process would displace remains an issue. The production of millions of tons of additional coke suitable for high-value applications will likely significantly reduce prices, which would increase overall coke consumption and decrease petroleum coke production for premium applications. One consequence could be greater use of petroleum coke in combustion applications, such as electric power generation. To the extent that additional petroleum coke is used for fuel applications, the net effect at the national level will be displacement of coal by coke. Another option is for refiners to gasify the excess petroleum coke to meet increasing hydrogen demand within refineries. To the extent that gasification of petroleum coke displaces petroleum or natural gas reforming, the net effect will be a small reduction in petroleum demand or natural gas demand.

---

[13] Information on the size and nature of the needle coke market is closely held and not available in the open literature.

[14] Paul J. Ellis and Christopher A. Paul, "Tutorial: Delayed Coking Fundamentals," Great Lakes Carbon Corporation, Port Arthur, Texas, 1998.

Continued Process Development and Fuel Testing

The development of the coal-tar blending process has advanced to the pilot-plant stage. Ten drums (55 gallons) of a prototype JP-900 fuel were produced in 2004. Most of this fuel was shipped to the Air Force Research Laboratory for testing, including small-scale turbine tests, and for analysis of combustion properties. Penn State is currently arranging for the operation of a pilot plant that can produce 1 barrel per day. This pilot plant would be operated under a subcontract to Intertek PARC at Intertek's Harmarville, Pennsylvania facilities.[15] The subcontract also includes purchasing and installing a new reactor at Harmarville. Current plans call for the production of 100 barrels of the prototype JP-900 fuel for testing, which would be conducted by or on behalf of the Air Force Research Laboratory.

Given the limited fuel production potential of the Penn State coal-tar blending process, we cannot identify any benefit for the Air Force or the U.S. government from proceeding with process development, especially the production of any fuel for fuel characterization or engine testing. Our judgment is that any further development of the coal-tar blending process will divert not only funds but also highly trained personnel, including Air Force Research Laboratory staff and the Penn State research team, from more-productive endeavors. Since engine testing is not productive, the delivery of 100 barrels of the prototype JP-900 fuel would also present a waste disposal problem for the Air Force Research Laboratory.

We have also considered whether pilot-plant operations based on the coal-tar blending process might be relevant for other processes Penn State is investigating for producing coal-derived jet fuels, such as the co-coking process. Given the current state of development of these alternatives, it is premature to assume that an economically viable process can be developed. It is, therefore, inappropriate to invest in expensive applied research directed toward upgrading and testing the products of such processes. Further, it is highly questionable, if not improbable, that information collected via the pilot plant (including subsequent testing) would contribute to the development or evaluation of such other approaches for producing coal-derived jet fuels.

The continuity of demand for premium coke over the longer term, beyond the next 10 to 15 years, is also in question. In particular, multiple research and development efforts are currently being directed toward finding a noncarbon substitute for the anode (sponge) coke used in aluminum production or, more broadly, for alternatives to the Hall-Héroult process. These efforts are driven by the potential for significant economic and environmental benefits, in particular, the potential to significantly lower carbon dioxide emissions associated with aluminum manufacture.

We recognize that the Penn State co-coking process is in the relatively early stages of development and that the mass balance shown in Figure 1 is approximate.

From our meeting with the Penn State research team and our review of their publications related to coal pyrolysis, it is our judgment that Professor Schobert and his colleagues have

---

[15] Intertek Group PLC is a London, UK-based firm specializing in testing, inspection, and certification of products, commodities, and systems. In 2005, Intertek purchased PARC Technical Services, Inc. PARC derives from the Gulf Oil Research and Development Company and was established as a separate entity in 1986, following Chevron's purchase of the Gulf Oil Corporation.

developed a center of excellence in coal pyrolysis at Penn State. We also believe that there is value in research directed toward liquids production via pyrolysis of coal or biomass or a combination of both coal and biomass. In particular, pyrolysis affords an energy-efficient means of converting coal or biomass to liquids. Rather than attempting to integrate coal pyrolysis into current oil refining operations, we suggest research that examines the opportunities of integrating pyrolysis with Fischer-Tropsch concepts for liquid-fuel production from coal and/ or biomass and with fermentation approaches for biomass conversion. In such cases, the co-product coke or char could be used as a gasifier feedstock for the production of synthesis gas or combusted to provide process heat.

It may be useful to pursue production of a new fuel additive that would provide better thermal stability or combustion properties than those of products that are currently available or that could be produced with state-of-the-art methods. However, if the Air Force requires a fuel additive with properties that are currently unavailable, we suggest that research directed toward such an additive consider the full scope of alternatives, including petroleum-based sources. A coal-based fuel additive for JP-8 will not displace significant amounts of conventional petroleum.

In conducting Air Force-sponsored research, the Penn State research team has focused most of its attention on producing a single product that could improve on JP-8. However, both processes we reviewed produce fluids with a mixture of constituents that, after further refining, will yield a broad slate of fuels, including gasoline and middle distillates. The same, of course, is true of the product slate produced by a modern petroleum refinery. To the extent that the Air Force continues to sponsor the development of advanced systems to produce jet fuel, that research should include consideration of the full product slate the advanced system would provide. In particular, fuel characterization and testing should address each of the major fuels that would be produced.

CHAPTER THREE

Findings and Recommendations

Finding 1

Our review of Penn State's work on JP-900 revealed a research team with considerable expertise in coal pyrolysis. Coal pyrolysis as a means of producing liquid fuels was studied extensively during the first half of the 20th century, but few research teams today have the expertise to address how pyrolysis might be exploited and combined with other liquid-fuel production approaches, including coal and biomass liquefaction via Fischer-Tropsch synthesis and other biomass routes to liquid-fuel production.

Recommendation

Consider supporting laboratory research and engineering analyses focusing on identifying possible opportunities by which pyrolysis can significantly improve the energy efficiency and costs of producing liquid fuels from coal and/or biomass.

Finding 2

The limited availability of coal tar seriously impedes the ultimate production potential of the Penn State coal-tar blending process. At most, successful development would produce only a few thousand barrels per day of jet fuel. The net impact on oil imports would be even smaller.

Recommendation

Cease all research directed toward developing this process, and cease testing fuels produced from this process. This includes terminating the planned pilot-scale operations at the Harmarville, Pennsylvania research site.

Finding 3

For the Penn State co-coking process, the limited availability of decant oil and the limited markets and high quality specifications for premium coke will limit liquid-fuel production to less than 140,000 barrels per day, only a portion of which would likely be suitable for use as a jet fuel. Less than 8,000 barrels per day of this production would be attributed to coal. The net increase in U.S. coal production due to developing the co-coking process would be negligible, about 2 million tons per year, and this estimate assumes that process economics are favorable and that one-half of U.S. decant oil production can be diverted to co-coking.

Recommendation

Cease all research directed toward developing (including product testing) any co-coking process concept that depends on large amounts of decant oil for the production of jet fuel. If work is to continue on co-coking, it should be limited to fundamental research investigating the feasibility of co-coking concepts that use feedstocks that are at least an order of magnitude more abundant than decant oil and that produce higher liquid and lower coke yields.

References

Baron, John T., Charles E. Kraynik, and Robert H. Wombles, "Strategies for a Declining North American Coal Tar Supply," in M. Sahoo and C. Fradet, eds., *Light Metals 1998 Métaux Légers*, Montréal, Que.: The Metallurgical Society of Canadian Institute of Mining, Metallurgy and Petroleum, August 1998.

Clifford, Carolyn, Energy Institute, Pennsylvania State University, personal communication, November 2006.

Ellis, Paul J., and Christopher A. Paul, "Tutorial: Delayed Coking Fundamentals," Port Arthur, Tex: Great Lakes Carbon Corporation, 1998.

Energy Information Administration, *Annual Energy Review 2005*, Washington, D.C.: U.S. Department of Energy, DOE/EIA-0384(2005), 2006a.

————, *Petroleum Supply Annual 2005*, Washington, D.C.: U.S. Department of Energy, DOE/EIA-0340(2005)/1, 2006b.

Fitzgerald, Kevin J., Vice President and General Manager, Carbon Materials and Chemicals, Koppers, Inc., personal communication, November 3, 2006.

Guercio, Vincent J., "Slurry Oils Supply and Demand," paper delivered at the 2004 Fuel Oil/Energy Buyers Conference, Miami Beach, Fla., CTC International, October 2004.

————, petrochemical consultant, CTC International, personal communication, January 10, 2007.

Peterson, D. J., and Sergej Mahnovski, *New Forces at Work in Refining: Industry Views of Critical Business and Operations Trends*, Santa Monica, Calif.: RAND Corporation, MR-1707-NETL, 2003. As of January 25, 2007:
http://www.rand.org/pubs/monograph_reports/MR1707/

Schobert, Harold H., Energy Institute, Pennsylvania State University, letter, December 12, 2006.